SO-BAH-691

DISCARD

1/05

WI

Let's Visit Canada

The Metric System

Joanne Mattern

PowerMath™

The Rosen Publishing Group's
PowerKids Press™
New York

Published in 2004 by The Rosen Publishing Group, Inc.
29 East 21st Street, New York, NY 10010

Book Design: Michael Tsanis

Photo Credits: Cover (landscape) © PhotoDisc; cover (Canadian flag) © EyeWire; p. 9 (CN Tower) © Alan Schein Photography/Corbis; p. 9 (observation deck) © Paul A. Souders/Corbis; p. 10 © Corbis; p. 13 © Grafton Marshall Smith/Corbis; p. 14 © Gunter Marx Photography/Corbis; p. 17 by Michael Tsanis; pp. 18 (Prince Edward Island), 21 (Stanley Park) © Peter Gridley/Taxi; p. 18 (Confederation Bridge) © Jan Butchofsky-Houser/Corbis; p. 20 (redwood tree) © V.C.L./Taxi.

Library of Congress Cataloging-in-Publication Data

Mattern, Joanne, 1963-
  Let's visit Canada : the metric system / Joanne Mattern.
       v. cm. — (PowerMath)
Includes index.
Contents: Where is Canada? — The metric system — Length and height — Measuring distance — Metric temperatures - How heavy? — Metric volume — Islands and a long bridge — Traveling west — Leaving Canada.
  ISBN 0-8239-8967-4 (lib. bdg.)
  ISBN 0-8239-8872-4 (pbk.)
  ISBN 0-8239-7380-8 (6-pack)
  1. Metric system—Juvenile literature. 2. Canada—Description and travel—Juvenile literature. [1. Metric system. 2. Weights and measures. 3. Canada—Description and travel.] I. Title. II. Series.
  QC92.5.M37 2004
  530.8'12—dc21
                            2003001257

Manufactured in the United States of America

Contents

Where Is Canada?

My family took a trip to Canada last year. Before we left, my mother showed me Canada on a map. Canada is part of the **continent** of North America. It is the country that is north of the United States.

My father said that Canadians use a different system of **measurements** from the ones we use in the United States. They use the metric system. The metric system doesn't use yards, ounces, or quarts. Instead, the metric system uses meters, grams, and liters.

My father told me that most countries around the world use the metric system. The United States is one of the few countries that doesn't use it. This book will teach you more about the metric system.

United States

Canada

United States

Mexico

South America

The Metric System

length	1 meter = 100 centimeters
	1 centimeter = 10 millimeters
distance	1 kilometer = 1,000 meters
weight	1 kilogram = 1,000 grams
volume	1 liter = 1,000 milliliters
temperature	0° Celsius = freezing point of water
	100° Celsius = boiling point of water

"Kilo" comes from the Greek word "*khilioi,*" which means "thousand." "Centi" comes from the Latin word "*centum,*" which means "hundred." "Milli" comes from the Latin word "*mille,*" which means "thousand."

The Metric System

Everything in the metric system is based on **units** of 10. The metric system uses meters to measure length. One meter, which is a little longer than a yard, is 100 centimeters long. One centimeter is 10 millimeters long. That means that a meter is 1,000 millimeters long.

The metric system uses liters to measure **volume**. One liter is a little less than 1 quart. There are 1,000 milliliters in 1 liter. The metric system uses kilograms to measure weight. One kilogram is about the same as 2.2 pounds. There are 1,000 grams in 1 kilogram.

Length and Height

The first place we visited was the city of Toronto in the **province** of Ontario. My favorite place in Toronto was the CN Tower. Dad said the CN Tower is the tallest building in Canada. We can use meters and centimeters to measure the **height** of the CN Tower. Mom told me that the CN Tower is 533 meters tall. To find out how tall the CN Tower is in centimeters, multiply 553 meters by 100 to get 53,300 centimeters.

$$
\begin{array}{r}
533 \text{ meters} \\
\times\ 100 \text{ centimeters per meter} \\
\hline
000 \\
0\ 00 \\
+\ 53\ 3 \\
\hline
53,300 \text{ centimeters}
\end{array}
$$

CN Tower

The CN Tower in Toronto has a glass floor 342 meters (or 34,200 centimeters) above the ground. It is strong enough to hold the weight of 14 hippos!

450 kilometers

Montreal

Toronto

Montreal

Measuring Distance

After we left Toronto, we drove to Montreal in the province of Quebec. The city of Montreal was named after the mountain around which it was built—Mount Royal. Mount Royal is 233 meters tall.

Kilometers measure long **distances**. A kilometer is 1,000 meters long. Montreal is 450 kilometers from Toronto. To figure out how many meters that is, I multiplied the number of meters in 1 kilometer (1,000) by the distance in kilometers (450). I wouldn't want to walk that far!

```
   1,000 meters per kilometer
 x 450 kilometers
 ─────────────────
   0 000
   50 00
 + 400 0
 ─────────────────
   450,000 meters
```

11

Metric Temperatures

We walked around the city of Montreal and saw many beautiful buildings. It was so warm that I took off my jacket. I saw a **thermometer** that said it was 20 **degrees**, which can also be written as 20°. I knew it had to be warmer than that.

Then I realized that metric **temperatures** are measured in degrees centigrade, or Celsius, which is different from the temperature scale we use in the United States. Twenty degrees Celsius is about 68° Fahrenheit.

Metric temperatures are based on units of 10, too. Zero degrees Celsius is the temperature at which water turns to ice. One hundred degrees Celsius is the temperature at which water boils.

°F °C

210 — — 100
200 — — 90
190 —
180 — — 80
170 —
160 — — 70
150 —
140 — — 60
130 —
120 — — 50
110 —
100 — — 40
90 — — 30
80 —
70 — — 20
60 —
50 — — 10
40 —
30 — — 0
20 —
10 — — -10
0 —
-10 — — -20
-20 — — -30
-30 —
-40 — — -40

100 degrees Celsius = boiling point of water

0 degrees Celsius = freezing point of water

Old Montreal

13

How Heavy?

We got hungry while we were walking around Montreal, so we stopped to buy some food. Then we sat outside by the water to eat. The food was measured in metric units, too. The weight of our box of crackers was listed in grams.

In the United States, smaller units called ounces add up to larger units called pounds. In Canada, grams add up to kilograms. There are 1,000 grams in 1 kilogram.

How many grams are in 5 kilograms? To find out, multiply 1,000 grams by 5 kilograms.

1,000 grams per kilogram
x   5 kilograms
―――――――――――
5,000 grams

There are
5,000 grams in
5 kilograms.

Metric Volume

After I had eaten my crackers I was very thirsty. Mom bought a bottle of juice. The bottle looked about the same size as a quart bottle, but the label said "1 liter."

A liter is the metric measurement for volume. A liter is slightly smaller than a quart. There are 1,000 milliliters in 1 liter.

If there are 1,000 milliliters in 1 liter, how many milliliters are in 10 liters? To find out, multiply 1,000 milliliters by 10 liters.

$$
\begin{array}{r}
1{,}000 \text{ milliliters per liter} \\
\times \quad 10 \text{ liters} \\
\hline
0\ 000 \\
+\ 10\ 00 \\
\hline
10{,}000 \text{ milliliters}
\end{array}
$$

There are 10,000 milliliters in 10 liters.

100% PURE

orange
juice

1 LITER

Prince Edward Island

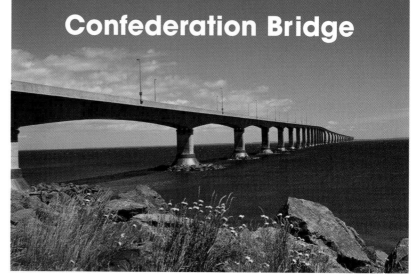

Confederation Bridge

An Island and a Long Bridge

After we left Montreal, we drove east to the Atlantic Provinces, which include Newfoundland, New Brunswick, Nova Scotia, and Prince Edward Island.

To get to Prince Edward Island, we drove across a long bridge called Confederation Bridge. It measured 13 kilometers long. To find out how many meters that is, we can multiply the number of meters in a kilometer (1,000) by 13 kilometers.

$$
\begin{array}{r}
\textbf{1,000 meters per kilometer} \\
\times \quad \textbf{13 kilometers} \\
\hline
\textbf{3 000} \\
+ \textbf{10 00} \\
\hline
\textbf{13,000 meters}
\end{array}
$$

**The bridge is
13,000 meters long!**

19

Traveling West

We left Prince Edward Island and drove west. We went to Vancouver in the province of British Columbia. We drove about 4,400 kilometers to get there! Vancouver was an exciting city. We visited a huge park called Stanley Park. My father said it is one of the biggest city parks in North America.

Stanley Park

Redwood trees grow in the mountains near Vancouver. They can grow up to 90 meters tall! How tall is that in centimeters? Multiply the number of centimeters in 1 meter (100) by 90 meters to find the answer.

Redwood Tree

100 centimeters per meter
x 90 meters
000
+ 9 00
9,000 centimeters

Leaving Canada

It was time to go home to the United States. Our plane flew home at 10,700 meters up in the air. How many kilometers is 10,700 meters? To find out, divide 10,700 meters by 1,000 meters per kilometer. The answer is 10.7 kilometers.

I had fun on our trip to Canada. I won't forget all the exciting things I saw there. I won't forget all the new metric measurements I learned, either!

$$
\begin{array}{r}
\underline{}\ \ \text{10.7 kilometers} \\
1{,}000\ \overline{)\ 10{,}700.0} \\
-10\ 00 \\
\hline
700 \\
-0 \\
\hline
7000 \\
-7000 \\
\hline
0
\end{array}
$$

22

Glossary

continent (KAHN-tuhn-uhnt) One of the 7 great masses of land on Earth.

degree (dih-GREE) A unit that measures how hot or cold something is.

distance (DIS-tuhns) The amount of space between two places.

height (HYT) How tall or high someone or something is.

measurement (MEH-zuhr-muhnt) A size or amount.

province (PRAH-vuhns) The name used for one of the main sections in some countries. Provinces are similar to states.

temperature (TEM-puhr-chur) A measure of how hot or cold something is.

thermometer (thur-MAH-muh-tuhr) A tool that tells how hot or cold the weather is.

unit (YOO-nuht) A standard amount by which things are measured.

volume (VAHL-yuhm) The amount of space inside something like a bottle, jar, or box.

Index

24